T0155599

Cambridge Elements ≡

Elements of Paleontology

STUDENT-CENTERED TEACHING IN PALEONTOLOGY AND GEOSCIENCE CLASSROOMS

Robyn Mieko Dahl
Western Washington University

CAMBRIDGE
UNIVERSITY PRESS

University Printing House, Cambridge CB2 8BS, United Kingdom

One Liberty Plaza, 20th Floor, New York, NY 10006, USA

477 Williamstown Road, Port Melbourne, VIC 3207, Australia

314–321, 3rd Floor, Plot 3, Splendor Forum, Jasola District Centre,
New Delhi – 110025, India

79 Anson Road, #06–04/06, Singapore 079906

Cambridge University Press is part of the University of Cambridge.

It furthers the University's mission by disseminating knowledge in the pursuit of education, learning, and research at the highest international levels of excellence.

www.cambridge.org
Information on this title: www.cambridge.org/9781108717861
DOI: 10.1017/9781108681483

First published 2018

A catalogue record for this publication is available from the British Library.

ISBN 978-1-108-71786-1 Paperback
ISSN 2517-780X (online)
ISSN 2517-7796 (print)

Student-Centered Teaching in Paleontology and Geoscience Classrooms

Elements of Paleontology

DOI: 10.1017/9781108681483
First published online: October 2018

Robyn Mieko Dahl
Western Washington University

Abstract: Research on learning and cognition in geoscience education and other discipline-based education communities suggests that effective instruction should include three key components: (a) activation of students' prior knowledge on the subject, (b) an active learning pedagogy that allows students to address any existing misconceptions and then build a new understanding of the concept, and (c) metacognitive reflections that require students to evaluate their own learning processes during the lesson. This Element provides an overview of the research on student-centered pedagogy in introductory geoscience and paleontology courses and gives examples of these instructional approaches. Student-centered learning shifts the power and attention in a classroom from the instructor to the students. In a student-centered classroom, students are in control of their learning experience and the instructor functions primarily as a guide. Student-centered classrooms trade traditional lecture for conceptually oriented tasks, collaborative learning activities, new technology, inquiry-based learning, and metacognitive reflection.

Keywords: student-centered pedagogy, geoscience education, paleontology education

ISBNs: 9781108717861 (PB), 9781108681483 (OC)
ISSNs: 2517-780X (online), 2517-7796 (print)

Contents

1 Introduction

Students are drawn to paleontology courses by the promise of dinosaurs, mass extinctions, and survival of the fittest. They don't always begin with the understanding that there is much more to the history of life on Earth than *Tyrannosaurus rex* and meteorite impacts. Our job as paleontology educators is to guide students toward a deeper understanding of evolutionary processes, the history of life, and the nature of science. And while it is relatively easy to lecture and to hope that the drama and narrative of the history of life will keep students engaged, research shows that our students benefit most from student-centered, active-learning pedagogy (National Research Council, 1996, 2000a, 2000b; Ruiz-Primo et al., 2011; van der Hoeven Kraft et al., 2011; Mogk and Goodwin, 2012; Freeman et al., 2014).

Student-centered learning shifts the attention in a classroom from the instructor to the students. In a student-centered classroom, students are in control of their learning experience and the instructor functions primarily as a guide. Because students are building new understanding through activity and inquiry, student-centered classrooms trade traditional lecture for hands-on explorations of content and concepts. This can mimic true scientific inquiry, ranging from carefully guided activities to open-ended explorations in which students develop their own questions, hypotheses, tests, and conclusions.

Research on learning and cognition suggests that students learn better in student-centered classrooms, and that effective instruction should include three key components: (a) activation of students' prior knowledge on the subject, (b) active learning pedagogy that allows students to address existing misconceptions and build new understanding of the subject, and (c) metacognitive reflections that require students to evaluate their own learning processes during the lesson (Figure 1) (National Research Council, 2000a). Designing courses that incorporate all three of these components while also accounting for class size, audience, and other course constraints is challenging, but educators can draw from the existing pool of validated student-centered activities, units, and lesson plans.

As paleontology educators, our practice will benefit greatly from the body of research produced by our colleagues in the geoscience education research community. The purpose of this Element is to provide an overview of the research on student-centered active learning pedagogy in introductory geoscience and paleontology courses, and to provide examples of these instructional approaches. Two central questions guide this Element. How do people learn? And what is unique about geoscience education as compared to other science fields?

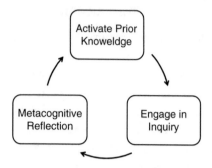

Figure 1 National Research Council framework for how people learn.
(A) Activate prior knowledge. (B) Build new understanding via active learning.
(C) Metacognition. (National Research Council, 2000a)

2 Student-Centered Learning

2.1 How Do People Learn?

Traditional, lecture-based courses often conflate teaching and learning. Lecturing on a topic delivers information to students, but there is little research support to show that students absorb, understand, and retain information that they receive through lecture (National Research Council, 2000a, 2000b). Thus, teaching does not necessarily result in learning in a lecture-based course. Lecture-based course design assumes that students enter the classroom as empty vessels, ready to be filled with new knowledge. In reality, students enter the classroom full of ideas, experiences, and preexisting frameworks for understanding the world around them. New knowledge must be delivered in ways that build onto students' preexisting frameworks, and lecture does not give students an opportunity to do that (National Research Council, 2000a).

While students in introductory courses are not likely to become true experts in the course topic, they should still develop the skills to think more like an expert. One important aspect of thinking like an expert is the ability to transfer understanding from one context to another (Hmelo-Silver et al., 2007). For example, a novice learner might successfully memorize the names and shapes of different types of trace fossils, but an expert would be able to classify new or unknown trace fossils based on their understanding of how traces are made. An expert has both a deep foundation of factual knowledge and the ability to fit new ideas into an existing framework (National Research Council, 2000a). An expert's frameworks for understanding are large and complex, and function like a filter to help sort and process new information (Hmelo-Silver and Nagarajan, 2002). With good course design, students can begin developing expert-like skills, even in introductory courses. Student-centered learning gives students this opportunity, which the science education research community refers to as *learning with understanding* or *deep learning*.

Students learn with understanding when they are able to engage authentically with course content (National Research Council, 2000a; Ruiz-Primo et al., 2011; Kyoungna et al., 2012; Freeman et al., 2014). Authentic engagement can take many forms, from activities in which students address real-world problems to true scientific inquiry. Inquiry, in this context, refers to activities that mimic the process of conducting scientific research. Inquiry in the classroom is often guided rather than open. In guided inquiry, students might follow a set of questions or procedures designed to prompt students to ask their own questions, pose hypotheses, design tests, and draw conclusions based on evidence. The results or conclusions of guided inquiry are often known or predictable to the instructor. Open inquiry is unstructured and less predictable. For example, students might be conducting true scientific research for which the results or conclusions are not known.

Student-centered classrooms are rooted in the theory that students will construct their own understanding of content through observation and inquiry during a lesson (Piaget, 1964; Fosnot, 2005). In a student-centered classroom, instructors function as guides as students engage in inquiry and problem-solving on their own or in small groups. Lectures are short and often occur midway through a lesson or even at the end. Lectures serve the purpose of clarifying or summarizing concepts that students have already explored earlier in the lesson.

The constructivist theory can prove difficult to put into practice, because it is hard to predict exactly what inferences or conclusions students will draw from their observations. This is in part because students will draw heavily from their prior experiences and knowledge when constructing new understanding (Piaget, 1964; National Research Council, 2000a; Fosnot, 2005). This prior knowledge is almost certainly incomplete and often includes misconceptions, so including opportunities for students to consider, write about, or discuss prior knowledge before engaging in new content allows the instructor to assess student's prior knowledge and address misconceptions (National Research Council, 2000a; Fosnot, 2005). Research also shows that if students' prior knowledge is not engaged, "they may fail to grasp the new concepts and information that are taught, or they may learn for the purposes of a test but revert to their preconceptions outside the classroom" (National Research Council, 2000a).

One way to elicit students' prior knowledge is a series of "initial ideas" questions designed to access students' prior knowledge or misconceptions at the beginning of an activity, and then check in with the class or individual groups to address any misconceptions before moving forward. Another approach to addressing misconceptions or confusion is to combine a constructivist approach with small interjections of explanatory reading or mini-lectures. Students can spend time engaging with the content and building their own understanding, and then they will have a more robust personal context in which to situate any subsequent reading or lecture.

Student-centered teaching benefits students, according to recent studies on the impact of student-centered teaching in geoscience classrooms (McConnell et al., 2005; Ruiz-Primo et al., 2011; Mogk and Goodwin, 2012; Freeman et al., 2014). For example, in a meta-analysis of research on the impact of student-centered instructional innovation in undergraduate science and engineering courses, Ruiz-Primo et al. (2011) found that students experience greater learning gains when student-centered instructional approaches were employed in large introductory science courses. In another meta-analysis of the impact of student-centered instruction, Freeman et al. (2014) found that these approaches improved student learning in all science, technology, engineering, and mathematics (STEM) disciplines, though the impact on geoscience courses was less pronounced than for other STEM disciplines. The sample size for geoscience courses was also much smaller than that for other STEM disciplines; only two geoscience studies met the requirements for inclusion in the analysis. This suggests an ongoing need for studies that quantitatively measure the impact of student-centered teaching in geoscience courses.

2.2 What Is Unique about Paleontology Education?

Some aspects of paleontology education are shared with other sciences, and some are unique to this field. Knowledge of this can help with instructional design, and a well-designed course will capitalize on paleontology's unique strengths while simultaneously addressing its unique challenges.

Student motivation for enrolling in paleontology and geoscience courses differs from student reasons for entering most other science fields. Many undergraduate students will have taken biology, chemistry, and physics courses in high school, and most STEM majors are required to take a year of chemistry and physics. Students will enroll in these courses because they are familiar subjects and are likely required. Paleontology and geoscience courses are not typically required for any majors besides geology. Furthermore, paleontology attracts students who are not STEM majors but are interested in learning about fossils. For these reasons, students in introductory geoscience or paleontology courses are likely there by choice but enter the course with less prior understanding of the content than students in other introductory science courses (McConnell et al., 2005; van der Hoeven Kraft et al., 2011).

Paleontology differs from most other sciences in that key phenomena cannot be directly observed. For example, even the most exquisitely preserved fossils are abstract representations of the living animal. Students are required to build mental models to translate what they see into what the fossil represents. While little research exists on students' conceptions of animals from fossils, we can extrapolate from research on students' spatial thinking skills and abstract reasoning to conclude that forming a mental model of a living animal is not an easy process (Chadwick, 1978; Kastens et al., 2009; Mogk and Goodwin, 2012). Evolution is another abstract concept that students encounter in paleontology courses. Evolution cannot easily be directly observed, so students have to build their own mental models in order to understand evolutionary change. For example, an instructor might illustrate the evolution of terrestrial tetrapods by showing students specimens or illustrations of several fossil fish, amphibians, and tetrapods. These specimens show the result of evolutionary change, but students must incorporate several concepts (natural selection, genetics, time, preservation, etc.) in order to build a mental model for the evolution of terrestrial tetrapods. Understanding the *process* is different and more difficult than memorizing the evolutionary progression of fossil taxa.

Paleontology and geosciences also often include opportunities to learn in the field. Class field trips can enrich students' learning experiences and lend themselves well to student-centered learning. In their synthesis of research on learning geosciences in the field, Mogk and Goodwin (2012) identified the primary benefits and opportunities of field experiences for geoscience students: embodiment, inscription, and initiation into a community of practice.

Embodiment is a concept borrowed from cognitive science that refers to our ability to use our physical relationship to an object or space to understand it in ways that are not possible for a machine or from a two-dimensional perspective. Frodeman (2003) describes this as "knowing your way around the topic, being oriented in conceptual space – or in an actual geographic and geologic space" (Frodeman, 2003, p. 127).

In the field, students can develop spatial thinking skills through the creation and use of inscriptions, which are constructed representations of natural phenomena like maps, sketches, and diagrams. Mogk and Goodwin (2012) refer to this as "inscription." Research has shown that this type of translation, from the real world to inscription, helps students filter and process information because they must distill the complexity of the real world down to only relevant concepts or details (Mogk and Goodwin, 2012).

Initiation into a community of practice means that in addition to learning how do geology, students learn how to *be geologists*. Mogk and Goodwin (2012) note that four categories of geoscience practice are learned in the field: language translated in the practice; tools used to acquire, organize, and advance community knowledge; shared ethics and values; and collective understanding of limits and uncertainties. Many students are initiated into the geoscience community in field camp, but they can gain these benefits during shorter field experiences too.

Field experiences can also present some challenges. Field trips are not possible for all courses or students; distance, time, expense, scale, safety, and accessibility all create barriers to field experiences. Furthermore, the real world is complex. If the activities and learning outcomes are not clear, students may have difficulty filtering and processing all the variables present at a field site (Ramasundaram et al., 2005; Mogk and Goodwin, 2012).

2.3 Student-Centered Instructional Design

Geoscience educators should approach course design with intention. Instructional design theory provides a framework to consider when building a new student-centered course or revising an existing one to incorporate more student-centered pedagogy. Using this framework ensures that a course addresses specific desired outcomes while also accounting for instructional conditions like class size, student demographics and cultural background, and time constraints (Reigeluth, 1999). The framework shown in Figure 2 helps guide an educator through design by first establishing specific desired learning outcomes for a course, module, class period or activity, then designing activities around those learning outcomes (Reigeluth, 1999). This approach, also known as "backward design," establishes the end goal and its appropriate

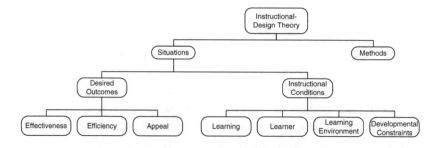

Figure 2 The components of instructional design theories (Reigeluth, 1999)

assessments (e.g., exam, lab report, essay) first and designs from that point back to the beginning of the activity, ensuring that every aspect of the activity helps to achieve the learning outcomes (Glossary of Education Reform, 2014).

Taking into account what we know about how people best learn paleontology and geoscience, we can now envision the ideal paleontology learning environment. The ideal course is one with clearly established learning goals and in which every activity and assignment is designed to guide students toward those learning goals. The ideal course is designed around the students and their learning experiences, and creates pathways for students to build new understanding of geoscience content and concepts. The ideal course positions students as scientists and allows them to ask their own questions and then to design their own methods for determining answers to those questions.

3 Student-Centered Pedagogy

Student-centered pedagogies are instructional methods that foreground student learning rather than teaching. Yacobucci (2012) provides a thorough list of active learning techniques, including some of the techniques described in this Element's activities, such as think-pair-share, concept maps, and guided inquiry. Additional techniques include peer teaching, brainstorming, responding to media reports, writing or drawing to learn, timelines, textual analysis case studies, games and role playing, debates and panel discussions, and service learning (Yacobucci, 2012). Ruiz-Primo et al. (2011) defined four broad categories of student-centered instruction: (1) conceptually oriented tasks, (2) collaborative learning activities, (3) use of technology, and (4) inquiry-based projects. These four categories, when used in concert with metacognitive reflection, encompass all aspects of the National Research Council's framework for how people learn.

3.1 Conceptually Oriented Tasks

Conceptually oriented tasks (COTs) are designed to engage students in conceptual schemes rather than isolated facts (Ruiz-Primo et al., 2011). COTs are intended to illustrate big ideas or to help students develop a framework for contextualizing new information. For example, students in a paleontology course must first understand the underlying process of evolution by natural selection before they can understand the driving forces behind the Mesozoic Marine Revolution or the Lilliput Effect. If students engaged in a COT focused on natural selection before they learned any information about specific evolutionary events, that COT would give them a framework to help relate, compare, or contrast specific evolutionary events.

COTs may also be used to at the beginning of an activity, module, or course to elicit students' level of understanding of or misconceptions about key concepts. For example, students may consider and discuss a series of questions or prompts before an activity begins in order to activate their prior knowledge of key concepts. Another example of a COT is a concept map, in which students visually represent the relationships that connect ideas, objects, or events (Novack, 1991; see Activity 1: Rock Cycle Concept Map for an example). COTs may engage students with real-world problems, which allows students to contextualize key concepts that can feel abstract or ungrounded until they are applied to a relevant issue. Ultimately, COTs focus on large-scale concepts, not isolated facts.

3.2 Collaborative Learning Activities

Collaborative learning activities are designed to engage students with peers in groups as small as pairs. Lessons designed around collaborative learning have been shown to promote a deeper understanding of concepts and content, because the sum knowledge of a group is greater than that of one individual (Lyle and Robinson, 2003; Tenney and Houck, 2003; Lorenzo et al., 2005; Arthurs and Templeton, 2009; Gilley and Clarkston, 2014; Bruno et al., 2017). Research on collaborative learning activities shows that group work improves learning for all students, not just those who are struggling with concepts (Gilley and Clarkston, 2014; Bruno et al., 2017). This is because collaborative learning gives students a space to discuss ideas and work through difficult concepts from multiple perspectives. Struggling students benefit from having a peer (rather than an instructor) explain concepts, and students who have a deeper under-standing of the concepts benefit from the opportunity to explain it or to teach it to their peers (Yuretich et al., 2001; Gilley and Clarkston, 2014; Bruno et al., 2017). Collaborative learning provides students with opportunities to engage in explanations and discussions as they describe their reasoning, interpretations, and solutions to problems (Ruiz-Primo et al., 2011).

A simple example of a collaborative learning activity is a think-pair-share activity, in which students consider a question or problem on their own (think), then discuss and refine their ideas with a partner (pair) before reporting back to the class (share). By allowing students to process on their own and with a peer before sharing, think-pair-share increases students' confidence in their ideas and stimulates discussion. Activity 2: Mystery Fossil Observation is based on the think-pair-share model.

Another example of a collaborative learning activity is the jigsaw activity (Figure 3). In a jigsaw, students begin the activity in a small "home group," then

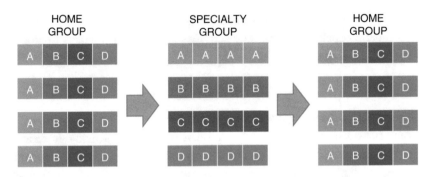

Figure 3 Jigsaw activity design, showing how students in home groups split apart into specialty groups, then return to home groups after mastering their specialty

split apart into "specialist groups" to focus on a specific aspect or topic. After mastering their "specialty," students return to their home group to tackle a complex problem collaboratively with representatives from each specialty. Jigsaw activities require a high degree of interdependence among students, because each specialist only possesses one "piece of the puzzle" and the whole group must work together to answer the question or to conduct the analyses (Aronson et al., 1978). Another benefit of the jigsaw activity is that it reduces the atmosphere of competition in a classroom; students are only successful when they work collaboratively with their group members (Aronson et al., 1978; Aronson and Patnoe, 2011). Activity 3: Sorting and Categorizing Fossils is an example of a jigsaw.

Student-centered classrooms may also utilize collaborative two-stage exams rather than traditional exams. In a two-stage exam, students take two identical versions of the same exam. In the first stage, they work independently. In the second stage, students work collaboratively in a small group (two to four students) on the same questions. Grades for two-stage exams are typically weighted, with ~75% assigned to the independent first stage and ~25% to the collaborative second stage. Studies of two-stage exams in large introductory geoscience courses show that students benefit greatly from the opportunity to work through the exam collaboratively with peers (Cortright et al., 2003; Lusk and Conklin, 2003; Gilley and Clarkston, 2014; Bruno et al., 2017).

Collaborative learning activities are central to a student-centered classroom. Working in collaboration and in small groups gives students the opportunity to engage directly with the content and with their peers, which promotes inquiry, discussion, and critical thinking skills. Even large lecture courses can be broken up into small groups throughout the class period to incorporate collaborative learning activities and to allow students to better process new content.

3.3 Technology

Technology can be incorporated into paleontology courses in a variety of ways, many of which are explored in other Elements in this paleontology series. In this context, technology can include any enhancement from viewing and analyzing digitized fossil specimens to building computer models to test hypotheses about evolutionary processes. Research shows an increase in student learning and engagement with content when technologies are incorporated into lessons.

3.3.1 Digitized Specimens

The ability to quickly and easily create high-resolution photographs and three-dimensional digital models of fossil specimens has revolutionized the study of paleontology, especially regarding the curation of collections. Most natural history museums are actively digitizing collections, and many are making their virtual collections available to researchers, educators, and even the general public. Digitized specimens can be used to enhance teaching collections, filling gaps and making the teaching of paleontology possible at institutions that do not have access to teaching collections.

3.3.2 Virtual Field Trips

As discussed earlier in this Element, field trips are a unique strength in geoscience education because learning in the field gives students the opportunity to both acquire knowledge and develop science skills and methods. Of course, field trips are not always logistically possible, especially for large enrollment courses, schools that are prohibitively far from "interesting" field sites, or students who are not physically able to conduct fieldwork. Virtual and augmented reality field trips are an effective alternative to actual field trips, and have been shown to increase student interest in studying geology because students can learn to "explore" a field site virtually without actually traveling there (McGreen and Sánchez, 2005; Bursztyn et al., 2017). For example, students who engage with virtual field trips scored higher on a geoscience interest survey, indicating that students were significantly more interested in learning geosciences than before taking the field trips

(Bursztyn et al., 2017). With new technology like connected classrooms and smartphone ubiquity, virtual and augmented reality field trips are becoming increasingly easy to use.

Virtual field trips can range from full virtual reality, in which students can walk around, explore, and interact with a field site, to limited virtual trips, in which students may examine a suite of photographs and other data collected from a locality. By allowing students to "travel" to field sites without leaving the actual classroom, virtual field trips can increase student interest in learning geosciences (McGreen and Sánchez, 2005; Bursztyn et al., 2017). For example, a virtual trip to the Grand Canyon is not equivalent to actually going to the Grand Canyon in person, but the virtual trip can activate student interest in it and increases their likelihood to attend actual field trips in the future (McGreen and Sánchez, 2005; Bursztyn et al., 2017). Programs like Interdisciplinary Teaching about Earth for a Sustainable Future (InTeGrate) and the Climate Literacy and Energy Awareness (CLEAN) Network provide peer-reviewed activities, modules, and lesson plans that incorporate computer models for geoscience educators to use in courses of all levels. InTeGrate and the CLEAN Network are both hosted by Carleton College's Science Education Resource Center (SERC, 2017).

3.3.3 Modeling and Scientific Databases

Modern methods in paleontology rely on modeling and database work, and the modern paleontology classroom should incorporate new methods into the curriculum. The Paleobiology Database and other large scientific databases provide an opportunity for students to learn modeling methods and conduct authentic research in the classroom (Uhen et al., 2016). Several classroom activities and ideas are posted on the Paleobiology Database website.

3.4 Inquiry-Based Projects

Inquiry-based projects are designed to trigger students' critical thinking, curiosity, and creativity. The degree of freedom given to students in inquiry-based projects can impact the learning outcomes. For example, in a guided inquiry, the instructor provides students with a set of preestablished questions for students to consider and the materials necessary for students to conduct the investigation. The instructor will guide students through the activity and the students will come to a predetermined conclusion. Free inquiry, which is much more like authentic scientific research, grants students much more independence. In truly free inquiry, students will design and conduct their own investigations and the instructor will have no way of knowing what conclusions students will draw. Activity 4: Introducing Inquiry and the Nature of Sciences uses guided inquiry to teach students about making observations and inferences, and provides students with the opportunity to pose and test hypotheses.

3.5 Metacognitive Activities

Metacognition includes students' self-reflection on knowledge and their ability to self-regulate their learning process (Figure 4) (Flavell, 1979; Schmitt and Newby, 1986; Schoenfeld, 1987). More generally, metacognition refers to "thinking about thinking," or a student's awareness of their own learning process. When students engage in metacognitive reflection, they need to consider three kinds of knowledge: declarative knowledge (knowing *what*), procedural knowledge (knowing *how*), and conditional knowledge (knowing *when* and *why*) (Smith and Newby, 1986). A scientist should recognize these types of knowledge as necessary components of scientific inquiry, so requiring students to engage in metacognitive reflection pushes students toward more authentic scientific inquiry. Metacognitive reflection can also improve a student's learning process by promoting better study skills and critical thinking (National Research Council, 2000a).

Incorporating metacognitive prompts in undergraduate geology courses can promote collaboration and improve learning outcomes (Spencer, 2017). A simple way to incorporate metacognitive reflection into a course is to incorporate metacognitive questions throughout an activity or module. High-level metacognitive questions include: (a) What are we trying to do? (b) Why are we trying to do that? (c) Are we making progress? (Schoenfeld, 1987; White et al., 2009). For example, if students are designing an experiment, ask them to explain why they chose to design the experiment the way they did. At the end of the activity, ask if their understanding of key concepts shifted over

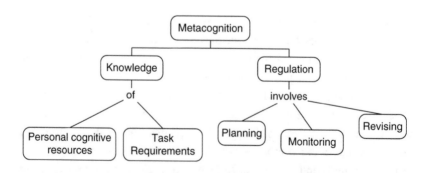

Figure 4 The components of metacognition (Schmitt and Newby, 1986)

the course of the activity, and ask them to describe the moments that triggered the change in understanding.

Journaling provides students with another opportunity to engage in meta-cognitive reflection by encouraging students to think about both what they learned and how they learned it. In a study of the impact of learning journals, Park (2003) found a range of positive outcomes for students, including feelings of ownership over their learning process, increased awareness of learning, increased self-confidence, increased engagement with the course material, and increased reflectivity.

4 Activity Examples

The following are examples of student-centered activities that can be used in paleontology or introductory geoscience courses. They emphasize inquiry, collaborative learning, and metacognitive reflection.

4.1 Activity 1: Rock Cycle Concept Map

In this activity, students use a suite of 20 rock specimens to build a rock cycle concept map. The final project is a visual representation of complex ideas and is intended to promote meaningful learning and to encourage students to think creatively about complex relationships (Novack, 1991). Concept maps have two key components: (a) concepts, which can be ideas, objects, or events; and (b) linking phrases, which describe the relationship between concepts (Novack, 1991). This activity gives students practice identifying rock specimens and asks them to build on their understanding of the rock cycle by linking each specimen to another via geological processes. Students should work in pairs or small groups. Figure 5 includes the instructions given to students and an example for each step.

Step 1

Examine rock specimens A–R. Complete the table by: (a) identifying the rock; (b) determining if it is an igneous, sedimentary, or metamorphic rock; (c) identifying major mineral constituents; and (d) describing the location of formation.

Step 2

Create a concept map showing how each rock was derived from other pre-viously existing rocks or materials. Each rock must be linked to the next by a geologic process. Number each process (1 through *n*). You may add rocks or materials onto the concept map (i.e., "sediments" or "magma").

Sample	Rock Name	Ig/Sed/Met	Major Minerals	Location of Formation
Y	Basalt	Igneous	Plagioclase, pyroxene	Mid-Ocean Ridge
Z	Blueschist	Met.	Glaucophane	Subducted Ocean Crust

Y. BASALT → 1. High Pressure Metamorphism → **Z. BLUE-SCHIST**

Process	Description
1. High pressure metamorphism	As ocean crust is subducted, it moves from a low temperature, low-pressure environment to a very high pressure environment. The temperature stays relatively low because it takes a long time to heat the subducted oceanic slab. The change in pressure causes plagioclase and pyroxene in the basalt to transform into glaucophane (and additional minerals) in blue schist. This is a metamorphic reaction, meaning it takes place in the solid state (no melting involved).

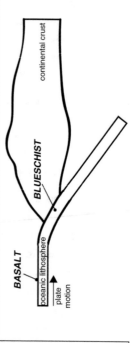

Instructions:

1. Examine rock specimens A-R and copmlete the table.

2. Create a concept map showing how each rock was derived from previously existing rocks (or other materials). Each rock must be linked to the next by a geologic process. Number each process (1 through *n*). You may add additional rocks or other materials onto the concept map (i.e., "sediments" or "magma").

3. Describe the geologic processes you invoked in your concept map. Consider these questions:
  a. What drives the process?
  b. How did conditions change between the "parent" rock and its product?
  c. How were minerals and other constituents affected by the process?
  d. Is the process closely connected to any other processes occuring at the same time?

4. Draw tectonic cross-sections to illustrate where in the Earth system each of the rocks (A-R) might have formed.

Figure 5 Activity 1: Rock Cycle Concept Map instructions and examples provided to students

Step 3

Describe the geologic processes you invoked in your concept map. Consider these questions:

a. What drives the process?
b. How did conditions change between the parent rock and its product?
c. How were minerals and other constituents affected by the process?
d. Is the process closely related to or influenced by any other processes occurring at the same time?

Step 4

Draw tectonic cross-sections to illustrate where in the Earth system each of the rocks (A–R) might have formed.

Step 5

Student pairs or groups share their final products (the concept map and tectonic cross-sections) with the rest of the class. The class can then discuss similarities and differences, with the instructor acting as a guide and providing clarification when needed. It should become obvious during discussion that there are many different ways to build an accurate concept map. Analyzing different models as a group gives students the opportunity to engage in metacognition by evaluating the strengths and weaknesses of their own model.

Assessment

Because students self-evaluate during the final discussion, they should be assessed primarily on engagement and participation. Content knowledge may be assessed later during a unit or final exam.

4.2 Activity 2: Mystery Fossil Observations

In this activity, students use fossils to practice making observations and inferences. This activity works best on the first day of a historical geology or introduction to paleontology course, in which students have likely never handled fossil specimens before. In addition to giving students practice and familiarizing them with fossil specimens, it gives the instructor an opportunity to evaluate students' prior knowledge. This activity also employs the think-pair-share model, in which students work on their own, share in a small group, and then discuss with the whole class. Think-pair-share gives students time to develop their own ideas, evaluate and refine those ideas with peers, and then present polished ideas to a large group.

To begin this activity, the instructor will distribute "mystery fossil" specimens to student pairs or small groups. This activity works best when the specimens include a range of invertebrate, vertebrate, plant, and trace fossil specimens.

Step 1

Specimen analysis. Answer the following questions on your own before discussing with your partner/group.

1. What is a fossil?
2. How do fossils form?
3. Make three observations about the specimen on your table.
4. Use your observations to make an inference about the specimen.
5. Discuss Questions 1–4 with your partner/group, then prepare a whiteboard to share your conclusions with the class.

This component of the activity does not need to be collected and graded by the instructor. The observations, interpretations, and conclusions that students write down should be drawn out during class discussion later in the activity.

Step 2

Class sharing. Each pair/group will present its specimen to the rest of the class, explaining its observations and inferences. The class is encouraged to ask questions during each presentation, and the instructor may choose to ask questions that spur discussion or help clarify student interpretations. The discussion should not move on to the next group until the whole class is satisfied with the current pair/group's interpretations.

Step 3

Class discussion. The instructor will lead a class discussion on the students' experiences making observations and drawing inferences about fossil specimens. Important metacognitive questions include: (a) "Was it easy to draw conclusions about your specimen?" (b) "Did your interpretation of your specimen change as you listened to other groups present?" (c) "What remaining questions do you have about your specimen?"

Step 4

Archive. The instructor should photograph each whiteboard and collect student answers to the questions in Step 1. These can be referred to throughout the course, as relevant content is covered.

Assessment

Students should be assessed primarily on their participation and engagement. The content knowledge gained in this activity is secondary to the practice of making observations and inferences.

4.3 Activity 3: Sorting and Categorizing Fossils Jigsaw

In this activity, students use the jigsaw activity model to develop their own classification system for a suite of mollusk and brachiopod shells. This activity would be most effective for students who have little to no experience with invertebrate fossils. By the end of the activity, students should be able to correctly classify an unknown shell based on criteria and descriptions developed during the activity.

Materials and Setup

Before class, the instructor prepares a box of 5–10 specimens for each of four categories: Bivalvia, Gastropoda, Cephalopoda, and Brachiopoda. The specimens can be modern shells, fossils, or a mix of both. Label the boxes A–D. The instructor should also set aside 5–10 additional specimens of each category, which will be used to quiz students later in the activity.

Step 1

Have the students form groups of four. This is their "home group." Have each member of the home group pick a letter (A–D) to indicate their "specialty." Once they have picked a number, students will disburse from the home group to their specialty groups.

Step 2

In the specialty groups, have the students spend a few minutes examining the entire suite of specimens in their box. After examining the whole suite, students write short descriptions of each specimen. They can write the descriptions as a group or individually.

Step 3

Once the specialty groups are done describing each specimen, they will synthesize a "consensus description" for the specimens in their box. Instruct the students that their description should be universal enough to accurately describe each specimen in the box and specific enough to be used to identify specimens from their category.

Step 4

When all specialty groups are done with their consensus descriptions, students return to their home groups and share their consensus descriptions with their home group members. They may choose to share specimens from their category box to help teach their home group members about their category. After each home group member has shared, all students should have a consensus description for each of the four categories.

Step 5

The instructor will now present each home group with four new specimens. Students will use their descriptions to classify the specimens. The instructor should check in with all groups as they classify the new specimens to ensure that each member of the groups understands the classification system.

Step 6

The instructor will now share the names and scientific descriptions of the four categories *without indicating which category each box represents.* These descriptions can be on a handout or projected at the front of the class. Students should read through the descriptions and compare them to their own descriptions, and decide as a group which box corresponds to which category.

Assessment

As a final quiz or summative activity, students should be required to correctly classify a suite of new specimens. They should also be able to explain the reasoning they used in their classifications.

4.4 Activity 4: Introducing Inquiry and the Nature of Science

This activity is used in an introductory geoscience course to explore the nature of science through making observations, collecting evidence, and making inferences (adapted from National Academies Press, 1998, pp. 66–73). In Steps 1 and 2, students examine a numbered cube (with numbers 1, 3, 4, 5, and 6 visible) and are instructed to use their observations to predict what number is on the unseen bottom face of the cube (see Figure 6).

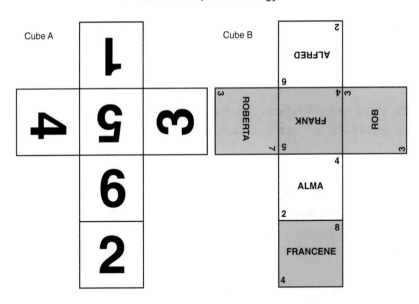

Figure 6 Activity 4: Introducing Inquiry and the Nature of Science, Cubes A and B (National Research Council, 1996)

In addition to a simple prediction, students are required to provide two clear lines of evidence that support their prediction. This provides an opportunity for students to develop their understanding of *observation* and *inference*. Through guided inquiry questions and discussion led by the instructor, students should conclude that inferences are made using observation and prior knowledge. In Steps 3–5 of the activity, students are given a second cube with a more complex set of markings on each face and asked again to predict what the bottom face of the cube looks like. In Step 6, students engage in whiteboard discussion about the activity. In this step, each small group will present its observations and final prediction about the bottom face of Cube B. The instructor can then lead a wider discussion about the nature of scientific inquiry, and should draw attention to the parallels between the bottom faces of the cubes and the Earth's interior. Humans have never traveled to the Earth's core, and yet we can predict with confidence what the characteristics of the core are, just as the students will never see the bottom face of either cube, and yet can make predictions with confidence about what the bottom faces look like.

Step 1

In a small group (three to four students), examine Cube A. You may rotate the cube to examine all sides, but do not lift the cube to look at the bottom face.

Write down your observations about the numbers and patterns that you observe on the sides of this cube.

Step 2

Describe how you interpreted these patterns to predict what number is on the bottom of Cube A. Describe at least two patterns that you used.

Step 3

Examine Cube B. Again, you may rotate the cube to closely examine all sides, but do not lift the cube to look at the bottom face. What is on the bottom of this cube? Be sure to record all of your observations and then propose an answer based on the patterns you observe in the exposed faces of Cube B.

Step 4

You may now design an experiment to obtain additional data about the bottom face of Cube B. You may use the small mirror provided to examine one corner of the bottom face. Use your observations of patterns to determine which corner you want to examine. Once you have decided which corner to peek at, slide that corner of the cube just far enough off the edge of the table to view one corner of the bottom face. Describe what you see.

Step 5

Discuss your observations with your group. Now what do you think is on the bottom face of the cube? Can you know for sure without lifting the cube? Once you have come to a group consensus, prepare a whiteboard to share with the rest of the class.

Step 6

Each group will share its predictions about the bottom of the cube. Your instructor will then lead a discussion about these predictions.

Assessment

Students should be assessed on their participation and engagement. Students are never shown the bottom of the cube, so they cannot be assessed on "correctness," only on process. The instructor may also assign an additional short writing assignment for students to summarize their newly formed understanding of the nature of science.

5 Conclusions

Creating a student-centered learning environment is challenging, but research shows how beneficial this pedagogy is for students. Student-centered activities take more time than lecturing, and the logistical difficulties of implementation increase with class size. Learning how to guide inquiry by showing rather than telling takes time and practice. Inquiry can lead students down unexpected lines of thinking, so an instructor in a student-centered classroom cannot always predict what conclusions students may come to and should be prepared to adapt. But a full transformation does not need to happen all at once. Instructors can make changes gradually by adding short activities and new student-centered modules over time. Instructors can draw from resources like this volume, and the Science Education Resource Center (SERC), which houses peer-reviewed geoscience activities, modules, lesson plans, and other resources related to student-centered learning (http://serc.carleton.edu). And after instructors gain comfort and confidence in student-centered teaching, they can implement larger-scale changes so that courses delineate specific learning outcomes and design concepts, lessons, and activities to correspond to each desired outcome.

References

Aronson, E., N. Blaney, C. Stephin, J. Sikes, and M. Snapp. (1978). *The Jigsaw Classroom*. Beverly Hills, CA: Sage Publishing Company.

Aronson, E., and S. Patnoe. (2011). *Cooperation in the Classroom: The Jigsaw Method*, 3rd edn. London: Printer and Martin, Ltd.

Arthurs, L., and A. Templeton. (2009). Coupled collaborative in-class activities and individual follow-up homework promote interactive engagement and improve student learning outcomes in a college-level environmental geology course. *Journal of Geoscience Education*, 57(5):356–371.

Bruno, B. C., J. Engels, G. Ito, J. Gillis-Davis, H. Dulai, G. Carter, C. Fletcher, and D. Bottjer-Wilson. (2017). Two-stage exams: A powerful tool for reducing the achievement gap in undergraduate oceanography and geology classes. *Oceanography*, 30(2):198–208.

Bursztyn, N., B. Shelton, A. Walker, and J. Pederson. (2017). Increasing undergraduate interest to learn geoscience with GPS-based augmented reality field trips on students' own smartphones. *GSA Today*, 27(6):4–10.

Chadwick, P. (1978). Some aspects of the development of geological thinking. *Journal of Geology Teaching*, 3:142–148.

Cortright, R. N., H. L. Collins, D. W. Rodenbaugh, and S. E. DiCarlo. (2003). Student retention of course content is improved by collaborative-group testing. *Advanced Physiological Education*, 27(3):102–108.

Flavell, J. H. (1979). Metacognition and cognitive monitoring: A new area of cognitive-developmental inquiry. *The Psychologist*, 34:906–911.

Fosnot, C. T. (2005). *Constructivism: Theory, Perspective and Practice*. New York, NY: Teachers College Press.

Freeman, S., S. L. Eddy, M. McDonough, M. K. Smith, N. Okoroafor, H. Jordt, and M. P. Wenderoth. (2014). Active learning increases student performance in science, engineering and mathematics. *PNAS*, 111:8410–8415.

Frodeman, R. (2003). *Geo-Logic: Breaking Ground between Philosophy and the Earth Sciences*. Albany, NY: State University of New York Press.

Glossary of Education Reform. (2014). Retrieved Dec. 3, 2017, http://edglossary.org/student-centered-learning/.

Gilley, B. H., and B. Clarkston. (2014). Collaborative testing: Evidence of learning in a controlled in-class study of undergraduate students. *Journal of College Science Teaching*, 43(3):83–91.

Hmelo-Silver, C. E., S. Marathe, and L. Liu. (2007). Fish swim, rocks sit, and lungs breathe: Expert-novice understanding of complex systems. *Journal of the Learning Sciences*, **16**(3):307–331.

Hmelo-Silver, C. E., and A. Nagarajan. (2002). "It's harder than we thought it would be": A comparative case study of expert-novice experimentation strategies. *Science Education*, **86**(2):219–243.

Kastens, K., C. A. Manduca, C. Cervato, R. Frodeman, C. Goodwin, L. S. Liben, D. W. Mogk, T. C. Spranger, N. A. Stillings, and S. Titus. (2009). How geoscientists think and learn. *EOS Transactions*, **90**:265–272.

Kyoungna, K., P. Sharma, S. M. Land, and K. P. Furlong. (2012). Effects of active learning on enhancing student critical thinking in an undergraduate general science course. *Innovative Higher Learning*, **38**(3):223–235.

Lorenzo, M., C. H. Crouch, and E. Mazur. (2005). Reducing the gender gap in the physics classroom. *American Journal of Physics*, **74**:118–122.

Lyle, K. S., and W. R. Robinson. (2003). A statistical evaluation: Peer-led team learning in an organic chemistry course. *Journal of Chemical Education*, **80**(2):121–124.

Lusk, M. and L. Conklin. (2003). Collaborative testing to promote learning. *Journal of Nursing Education*, **42**(3):121–124.

McConnell, D. A., D. N. Steer, K. D. Owens, and C. C. Knight. (2005). How students think: Implications for learning in introductory geoscience courses. *Journal of Geoscience Education*, **54**(4):462–470.

McGreen, N, and I. A. Sánchez. (2005). Mapping challenge: A case study in the use of mobile phones in collaborative, contextual learning. In P. Isaías, C. Borg, P. Commers, and P. Bonanno, eds., *Proceedings of the IADIS International Conference Mobile Learning*. Qawra, Malta: IADIS pp. 213–217.

Mogk, D. W., and C. Goodwin. (2012). Learning in the field: Synthesis of research on thinking and learning in the geosciences. *Geological Society of American Special Papers*, **486**:131–163.

National Academies of Sciences. (1998). Activities for teaching about evolution and the nature of science. In *Teaching about Evolution and the Nature of Science*. Washington, DC: National Academies Press, pp. 61–73.

National Research Council. (1996). *From Analysis to Action: Undergraduate Education in Science, Mathematics, Engineering, and Technology*. Washington, DC: National Academies Press.

National Research Council. (2000a). *How People Learn: Brain, Mind, Experience, and School*, expanded edn. Washington, DC: National Academies Press.

National Research Council. (2000b). *Inquiry and the National Science Education Standards: A Guide for Teaching and Learning*. Washington, DC: National Academics Press.

Novack, J. D. (1991). Clarify with concept maps: A tool for students and teachers alike. *The Science Teacher*, **58**:45–49.

Park, C. (2003). Engaging students in the learning process: The learning journal. *Journal of Geography in Higher Education*, **27**(2):183–199.

Piaget, J. (1964). Development and learning. *Journal of Research in Science Teaching*, **28**:213–224.

Ramasundaram, V., S. Grunwald, A. Mangeot, N. B. Comerford, and C. M. Bliss. (2005). Development of an environmental virtual field laboratory. *Computers and Education*, **45**:21–34.

Reigeluth, C. M. (1999). What is instructional-design theory and how it is changing? In C. M. Reigeluth, ed. *Instructional-Design Theories and Models, Volume II*. New York, NY: Routledge, pp. 5–30.

Ruiz-Primo, M. A., D. Briggs, H. Iverson, R. Talbot, and L. A. Shepard. (2011). Impact of undergraduate science course innovations on learning. *Science*, **331**(6022):1269–1270.

Schmitt, M. C., and T. J. Newby. (1986). Metacognition: Relevance to instruction design. *Journal of Instructional Design*, **9**(4):29–33.

Schoenfeld, A. H. (1987). What's all the fuss about metacognition? In A. H. Schoenfeld, ed., *Cognitive Science and Mathematics Education*. Hillsdale, NJ: Lawrence Erlbaum, pp. 189–215.

Science Education Resource Center. (2017). Retrieved Dec. 2, 2017, http://serc.carleton.edu.

Spencer, D. (2017). Enhancing Socially-Shared Metacognition in Introductory Geology. Unpublished PhD dissertation, North Carolina State University.

Tenney, A., and B. Houck. (2003). Peer-led team learning in introductory biology and chemistry courses: A parallel approach. *Journal of Mathematical Sciences*, **6**:11–20.

Uhen, M. D., L. Lukes, C. George, and R. Lockwood. (2016). Build creative thinking into the STEM undergraduate classroom experience using large data-bases: The Paleobiology Database example. *Innovations in Teaching and Learning Conference Proceedings Vol. 8*, http://dx.doi.org/10.13021/G84G7B.

van der Hoeven Kraft, K. J., L. Srogi, J. Husman, S. Semken, and M. Fuhrman. (2011). Engaging students to learn through the affective domain: A new framework for teaching in the geosciences. *Journal of Geoscience Education*, **59**:71–84.

White, B., J. Fredericksen, and A. Collins. (2009). The interplay of scientific inquiry and metacognition: More than a marriage of convenience.

In D. J. Hacker, J. Dunlosky, and A. C. Graesser, eds., *Handbook of Metacognition in Education*. New York, NY: Routledge, pp. 176–205.

Yacobucci, M. M. (2012). Using active learning strategies to promote deep learning in the undergraduate paleontology classroom. In M. M. Yacobucci and R. Lockwood, eds., *Teaching Paleontology in the 21st Century*. Paleontological Society Special Publication Vol. 12, pp. 135–153.

Yuretich, R. F., S. A. Kahn, R. M. Leckie, and J. J. Clement. (2001). Active-learning methods to improve student performance and scientific interest in a large introductory oceanography course. *Journal of Geoscience Education*, **49**(2):111–119.

Acknowledgments

Many thanks to my colleagues in the Science Math and Technology Education (SMATE) Program at Western Washington University for their ongoing commitment to and discussions about student-centered learning. This Element also benefited greatly from the comments and insight provided by my reviewers. Special thanks to Susan DiBari for her generous mentorship and to Jenny Marie Forysthe for her many discussions on student-centered learning and editing suggestions for this Element.

Cambridge Elements ≡

Elements of Paleontology

Editor-in-Chief

Colin D. Sumrall
University of Tennessee

About the Series

The Elements of Paleontology series is a publishing collaboration between the Paleontological Society and Cambridge University Press. The series covers the full spectrum of topics in paleontology and paleobiology, and related topics in the Earth and life sciences of interest to students and researchers of paleontology.

The Paleontological Society is an international nonprofit organization devoted exclusively to the science of paleontology: invertebrate and vertebrate paleontology, micropaleontology, and paleobotany. The Society's mission is to advance the study of the fossil record through scientific research, education, and advocacy. Its vision is to be a leading global advocate for understanding life's history and evolution. The Society has several membership categories, including regular, amateur/avocational, student, and retired. Members, representing some 40 countries, include professional paleontologists, academicians, science editors, Earth science teachers, museum specialists, undergraduate and graduate students, postdoctoral scholars, and amateur/avocational paleontologists.

Paleontological

S O C I E T Y

Cambridge Elements ☰

Elements of Paleontology

Elements in the Series

These Elements are contributions to the Paleontological Short Course on *Pedagogy and Technology in the Modern Paleontology Classroom* (organized by Phoebe A. Cohen, Rowan Lockwood, and Lisa Boush), convened at the Geological Society of America Annual Meeting in November 2018 (Indianapolis, Indiana USA).

Flipping the Paleontology Classroom: Benefits, Challenges, and Strategies
Matthew E. Clapham

Integrating Macrostrat and Rockd into Undergraduate Earth Science Teaching
Pheobe A. Cohen, Rowan Lockwood, and Shanan Peters

Student-Centered Teaching in Paleontology and Geoscience Classrooms
Robyn Mieko Dahl

Beyond Hands On: Incorporating Kinesthetic Learning in an Undergraduate Paleontology Class
David W. Goldsmith

Incorporating Research into Undergraduate Paleontology Courses: Or a Tale of 23,276 Mulinia
Patricia H. Kelley

Utilizing the Paleobiology Database to Provide Educational Opportunities for Undergraduates
Rowan Lockwood, Pheobe A. Cohen, Mark D. Uhen, and Katherine Ryker

Integrating Active Learning into Paleontology Classes
Alison N. Olcott

Dinosaurs: A Catalyst for Critical Thought
Darrin Pagnac

Confronting Prior Conceptions in Paleontology Courses
Margaret M. Yacobucci

The Neotoma Paleoecology Database: A Research Outreach Nexus
Simon J. Goring, Russell Graham, Shane Oeffler, Amy Myrbo, James S. Oliver, Carol Ormond, and John W. Williams

Equity, Culture, and Place in Teaching Paleontology: Student-Centered Pedagogy for Broadening Participation
Christy C. Visaggi

A full series listing is available at: www.cambridge.org/EPLY

Printed in the United States
By Bookmasters